# THE CALCULUS VIRGIN

## An Artist's View of the Language of Calculus

To Lynda
in the spirit of
humanistic
mathematics
Louis Leithold
A.P. Calculus Reading
Clemson 6/18/92

92 93 94 95 96 10 9 8 7 6 5 4 3 2 1

# THE CALCULUS VIRGIN

## AN ARTIST'S VIEW OF THE LANGUAGE OF CALCULUS

by

**d'ARCY HAYMAN**

with introduction and notes by

**LOUIS LEITHOLD**

Tortue Publications

# CONTENTS

THE AUTHORS

**d'Arcy Hayman** is an artist, poet, former faculty member of the University of California at Los Angeles and Columbia University, and for twenty years from 1960 through 1980 head of the International Arts Program for the United Nations in Paris.

**Louis Leithold** is the author of **The Calculus with Analytic Geometry**, first edition published in 1968 and sixth edition in 1990. This book has been among the leading calculus texts used in universities, colleges, and secondary school advanced placement programs for more than twenty years. Writing this and other mathematics textbooks, lecturing, and university teaching have comprised his professional life.

# THE CALCULUS VIRGIN

**Calculus**, in its original Latin context, was the word for pebble used in the ancient world for counting. In modern language the word **calculus** still refers to a stony mass or deposit that can be formed in the human body, especially in the kidney. So stones and counting are the basic references here. Of course the term as it is understood by most of us today refers to a method of mathematical analysis and calculation using symbols.

**Virgin**, also coming from the Latin, was a name of one of the signs of the zodiac, **Virgo**. While modern tongues often connect the word **virgin** with a young maiden it can also mean a pure, unexplored state, as well as the first or initial experience. In this sense I interpret the **virginal** act as being an awakening, an introduction to a new dimension of life.

I have named myself **the calculus virgin** in this book because the whole work stems from this artist's **virginal** encounter with **calculus**.

The drawing combines four symbols: the counting stones of past time, the zodiac where **Virgo** predominates, an imagined **Virgin** Goddess of an ancient culture, and the terminology of calculus.

d'Arcy Hayman

VIRGO
LEO
CANCER
GEM
LIBRA
SCORPIO
TARIUS
1 2 3 4 5 6 7 8 9 10
I II III IV V VI VII VIII IX X
$\frac{dy}{dx} = \frac{dy}{du} \cdot \frac{du}{dx}$
$\lim_{\Delta x \to 0} \sum_{i=1}^{n} f(c_i)\Delta x = \int_a^b f(x)\,dx$
$\ln x = \int_1^x \frac{1}{t}\,dt$
$\lim_{x \to 0} \frac{\sin x}{x} = 1$
$\int u\,dv = uv - \int v\,du$
$dz = \frac{\partial z}{\partial x}dx + \frac{\partial z}{\partial y}dy$
d'Arcy

# INTRODUCTION

In March of 1988, I conducted a seminar for teachers of advanced placement calculus that consisted of a discussion of the theory behind some important calculus topics as well as techniques of teaching them. My good friend d'Arcy Hayman attended this seminar as an observer. I did not encourage her presence because she had no background in mathematics. A description of d'Arcy's reaction to the language of calculus follows this introduction. This description was adapted by d'Arcy from an audio tape recording of what she told me the day after the seminar.

Nothing in my many years of teaching calculus had prepared me for this experience that she characterized as both passionate and thrilling. I was so excited about this revelation that the following month I read a transcription of the tape of d'Arcy's reaction as part of a lecture to a group of four hundred calculus teachers at the annual meeting of the National Council of Teachers of Mathematics in Chicago. Since then by popular demand I have repeated this lecture to other groups of calculus teachers throughout the country. The enthusiasm generated by these presentations and the many requests I received for copies of the text of this lecture persuaded d'Arcy and me to produce **The Calculus Virgin**.

d'Arcy's view of the language of calculus is reflected by fifty-three drawings and their accompanying texts. The first drawing, **The Calculus Virgin**, precedes this introduction. The second, which follows this introduction, was presented to me by d'Arcy the day after the seminar as a result of her looking through a booklet containing sample calculus problems. The next fourteen show images pertaining to vocabulary d'Arcy heard at my seminar, while the remaining thirty seven illustrate words she read while perusing a copy of my calculus book. On the top half of the page opposite each drawing, except for the first, I have written the mathematical text containing the word or group of words represented. On the bottom half of that same page appears d'Arcy's explanation of the symbolism of the drawing as well as cultural references related to the word or group of words represented.

Louis Leithold

## CALCULUS PROBLEMS

Excerpt from a page of calculus problems:

1. The normal to the curve represented by the equation $y = x^2 + 6x + 4$ at the point $(-2, -4)$ also intersects the curve at $x =$?
2. If $y = x + \sin(xy)$, then $\frac{dy}{dx} =$?

---

The visual image in this drawing is meant to make calculus people laugh. I drew it a few minutes after I was handed the first page of calculus problems I had ever seen. I combined the first line of problem number 1 with the first line of problem number 2. The key words or symbols for me were **normal**, **curve**, **sin**, **then *dy***, and the **question mark**. I read these lines as a morality poem. To illustrate these concepts I used the stereotypes and cliches that western cultures have employed to represent these ideas. For **normal**, I picture a man in business suit and tie which takes the form of a granite bust: solid; strong; a monument to normality. The **curve** can be seen as a rainbow, standing for hope on the horizon. The nude male figure with fig-leaf represents the original **sin** in Genesis, which among other things resulted in making humans ashamed of their bodies. **Then *dy*** is shown at the bottom with the face of a person in final repose. The **question mark** makes us wonder: Is death the final state?

The normal To the curve
represented by the equation
If y=x + sin (xy)
Then dy/dx.
d'Arcy

# THE ARTIST'S VIEW

d'Arcy Hayman speaking to Louis Leithold:

I was very grateful for the opportunity to attend your seminar on the teaching of calculus. I knew it would be an interesting experience even before I went. You asked me the question "What on earth could you possibly expect to enjoy or obtain from a subject in which you have neither background nor knowledge?" You only wanted to protect me from something that might be boring. And even though I certainly was aware of the fact that I had no mathematical background, a calculus virgin of sorts, I still was eager to see and hear this presentation. So I went with anticipation and was rewarded with a tremendous, even thrilling, experience.

I told you at that time, I was indeed the calculus virgin. But as happens sometimes with self-proclaimed virgins, there was more in my past than I have thus far admitted to you. And before we go on, I feel the need to confess. While it is true that I had no knowledge of calculus and almost no direct contact with mathematics in general, I must explain that even as a child I became aware of the great poetry and mystical quality of mathematics. I saw in early human history that the arts and sciences were entwined; they worked as one in the evolution and enhancement of civilization. In ancient cultures, such as found in Egypt, China, India, Greece, as well as in the Celtic and other Anglo Saxon cultures and those of the Aztec, Toltec, Mayan, Incan, and North American indigenous peoples, mathematics and the arts were closely knit resulting in some of the greatest monuments and artifacts the world has ever known. My intuition about the superb inner beauty of mathematics continued throughout my life. Because I was never fortunate enough to enter the world of numbers, I attempted to understand what I could through words by reading profound and revealing works such as Einstein's "Essays in Science." This might help you see why despite my lack of background in calculus, I wanted to attend your lecture on the subject.

You will laugh when you hear my notes and questions. Your laughter may be caused by my virginal innocence of your field, but beware, I too love to laugh and therefore I have seized every opportunity to invent a bit of humor along the way. I wrote notes, as I told you, but they are not the ones that you or your students or anyone with a knowledge of calculus would make. I do not want to convey the thought that in this one session of involvement with calculus I could truly enter its symbolism. Certainly I only could make associations with its literary content, and then bring to these words the cultural references that have meaning in my own life as an artist and world traveler. I fully realize that my references, both verbal and visual, might have little or nothing to do with the meanings they have in calculus. But perhaps they will bring you another view, which illustrates one of the wonderful things you say in calculus: "both sides exist."

My first note refers to your opening remarks. You asked teachers in attendance "What one topic is most important to you in your teaching, the one you would want first to bring to your students?" Several terms were offered by the group and you discussed the terms. But just the mention of these terms was fascinating to me and made me want to know so much more about each one.

First came the concept of

**limit.**

The mental image of **limit** immediately sparked my imagination because **limits** in my life as an artist are so important. **Limits** in all of life are basic, not only as restrictive concepts but as challenges. When you have **limits** you can work because working within **limits** is a structuring and ordering of life that you would not have if you did not have **limits.**

Then I put down

**optimization.**

**Optimization**? My mind went wild. What could that mean? The optimum means the greatest, the farthest reaches of anything or everything. To work with a concept of **optimization** sounded so marvelous to me.

The next term was of primary importance:

**intuition.**

**Intuition** struck me as central because my whole life is one that is based a great deal on **intuition**. For the visual artist, the composer of music, the poet, and all other persons in the arts, **intuition** plays a major role. It is good to be reminded of the importance of **intuition** in the sciences and mathematics. I remember the great thrill it brought me in my youth on learning that Einstein, in formulating his world changing theories and equations, first went through a series of visual and muscular sensations, **intuitions** about certain relationships, patterns, and connections. Only after this stage of conceptual musing did he begin to use the tools at his command, the language of mathematics, to give structure to his preliminary **intuition**. In this sense I see the artist and the scientist as one creative human spirit. Call it imagination, call it revelation, or call it **intuition**, it is at the heart of all discovery and invention.

Then another word stirred my blood. In Paris in 1964, I had a one-person exhibition of fifty huge paintings all based on one theme:

**origin.**

**Origin**, of course, is fundamental to every field: the **origin** of the universe, the **origin** of life, the **origin** of an idea.

And you said this group of words:

**oscillates forever about the origin**.

This terminology, a poem in itself, sent me into other realms of experience. I was soaring. The concept of **oscillating forever about the origin** is so exhilarating, I was overwhelmed.

You referred to

**a different domain**.

One can think of many things in relation to that. I can see that the artist's life is a **different domain**. One of my own paintings, entitled "Man Apart from the Crowd," symbolizes the importance of the individual, the unique, the different.

And then you said

**$x$ is getting closer and closer to one**.

Just the thought of getting **closer and closer to one** produces a superb sensation with ideas of unity, oneness, completion, a whole; it was a beautiful statement.

You used the word

**mysterious**.

At one time you purposely did not give some information then because, as you said, "I like to make it more **mysterious**." Mystery, mathematics, and art belong together.

Now here is one note that is circled because it is so provocative:

**an irrational exponent that makes sense**.

Well, that struck lightning in me. As you know, in our culture, we are constantly fighting the prejudice that favors the so called rational over any other mind-state such as dreaming, visions, intuition, or other forces that have been with us since the beginning of human history. But, of course, this separation of rational from **irrational** is an artificial one. The cliches of artist as dreamer, close to the **irrational**, and the mathematician as cold and calculating (pun), fully rational, are misinterpretations that have gained public acceptance in western cultures since the Renaissance. We only have to look back to Leonardo Da Vinci to see that genius of universal awareness includes the arts and sciences, the "**irrational**" and "rational." At last the public's perception of these matters soon will be changing as we begin to give attention to such new fields as "the science of chaos." I rejoice in hearing your reference in calculus to **an irrational exponent that makes sense**.

Then there is something here about mathematics before calculus, like algebra, geometry, and trigonometry. You said you could not define this exponent with all the other mathematics that precedes calculus. I took note of the cumulative building of ideas that progress from non-expressible to being expressible.

Another statement that fired my imagination:

**This function obeys**.

Now I did not hear what it **obeys**, after that or before that, but just the perception of a **function obeying** brought images of some medieval submissiveness.

The term

**intervals**

is inherent in music, art, poetry, and everything else in life.

The next group of words took me to my interest in eastern philosophy:

**both sides exist**.

I was reminded of Taoist and Buddhist concepts of opposites with the understanding that there are two equally important sides of things, each side complementing the other.

You spoke of

**integrating a power**

and

**perfectly rigorous**.

This brought a strong and virile image to mind that, for me, had an erotic flavor to it.

I was exalted when you said

**infinity**.

That is an exquisite notion. As we all grow older with the human awareness of the inevitability of physical death, **infinity** becomes an ever more profound concept. Since the earliest recordings of human concerns through art works and later through words and numbers, there has been unending interest in the connection between the human spirit and the **infinite**.

How I wish I could deeply know

**the language of calculus**

so that I could join you on your journey to "**infinity**."

When I reflect back on my first time in calculus at your seminar, I realize that technically I remain the calculus virgin, given the fact that the calculus message did not fully penetrate my mathematically untrained mind. But I was gloriously touched by the poetry of the experience. You have asked me to share my verbal and visual notes with you and I do this gladly in the hope that this exchange between artist and mathematician can bring new light and joy to both of us.

As I look over my notes on the seminar and the stream of thoughts that resulted, this occurs to me:

In our age of specialists and specializations, as we go further into more fragmented fields in life, what I feel we need now and in the future is to open the barriers, seek new vistas, broaden horizons and perspectives, find ways to link all disciplines, and see the universality in them so that we can be enriched by mathematics, science, the humanities, and the arts. All these human languages are important. Of course I realize that we do not have time in our brief stay on this planet to deeply emerse ourselves in each of these languages. But I hope that we can find a way in our educational structures to bring to all people not only the computational aspects of mathematics and facts, names, dates, and places in the history of art, but the magic of human creativity in all its forms. I think this will be important to our understanding of the universe and our place in it, to our well being, and to the future of civilization.

## LIMIT

The two fundamental mathematical operations in calculus are differentiation and integration. These operations involve the computation of the derivative and the definite integral, each of which is based on the notion of **limit**.

---

Life's **limits** are part of the ordering process. They give choice, direction, and challenge. They make possible all human action. The symbols I have chosen show the individual artist in the lower triangle in the process of deciding on the subject matter for a painting. At first he is confronted by everything in the physical and metaphysical universe, as shown on the two side panels. He has to begin to set **limits** for himself after he accepts the tangible **limits** of time, space, materials, and so forth. The first **limit** he sets is that he will concern himself with the realm of flowers, as pictured in the center panel. The next **limit** is to focus on one particular flower to represent in the painting, as shown in the square above the triangle. Of course, the artist must set thousands of **limits** on each step of the way to the finished work, as must every person in all aspects of life. The involvement of the human brain and heart as well as extra-terrestrial forces enter into these **limit**-setting functions.

d'Arcy

# OPTIMIZATION

The greatest function value on an interval is called the absolute maximum value and the smallest function value on an interval is called the absolute minimum value. These function values are used to solve **optimization** problems, those involving maxima and minima.

---

The lower hemisphere in the drawing is the realm of individuals who work and struggle to survive in a world of obstacles, challenges, disappointments, and in some cases defeat. The one individual in the center of this hemisphere reaches up and somehow transcends the ordinary, extends himself into the domain of possibility, and makes **optimum** use of his potential. The structure in the middle of the drawing represents a labyrinth through which this individual must travel with his intelligence, creativity, faith, and love before he can reach Nirvana, the enlightened state in the top sphere. The side bands are the sages and saints of all history who must judge him and then honor him in his new home of **optimal** fulfillment.

d'Arcy

## INTUITION

To arrive at a definition of the number $A$ that represents the area of a region $R$, our **intuition** tells us that $A$ must be at least as great as the area of any polygonal region contained in $R$, and it must be no greater than the area of any polygonal region containing $R$.

---

To illustrate the important role played by **intuition** in the lives and work of artists and scientists, I picture an individual who is making a connection with some vast source outside himself, represented by the large sphere. This person's body is marked by lines of the billions of energetic impulses that act as transmitters and receptors of universal communications. He carries the Hindu symbols of the points of enlightenment, one circle on his forehead and one close to his navel. He seeks the third circle that will complete the trinity of **intuition**. His hands search among the endless bands of **intuitive** stimuli until there appears just within his grasp the precious cosmic circle. We witness the miraculous birth of **intuition**.

d'Arcy

# OSCILLATES FOREVER ABOUT THE ORIGIN

The function defined by $f(x) = \sin(1/x)$ does not have a limit as $x$ approaches 0 from either the right or the left. The origin is not on the graph of this function but the graph **oscillates forever about the origin** from $-1$ to 1 as $x$ gets closer and closer to 0.

---

**Origin** is represented by the band across the top, which is a field of spiral eggs waiting for and being entered by the racing sperm lines, finally resulting in the human fetus below. The **oscillating** circles symbolize all of the elements that contribute to the formation of this infant: the parental genes; the stars; the sun; the moon; the planets; the culture to which he is born; the time and place.

d'Arcy

## A DIFFERENT DOMAIN

If $f(x) = \sqrt{x}$, the function having values $f(x^2)$ has **a different domain** than the function having values $[f(x)]^2$.

---

I have invented three **domains** each of which exists separate from the others even though they are closely connected and interact in many ways, as indicated by the lines between the suggested spheres. The angel in the dress of some ancient eastern culture represents the spiritual **domain**. The sphere on the right is the land of flora and fauna. The sphere on the left is the human **domain**, the world of human invention and the realm of civilization. Each of these **domains** contributes to and gains from the others. They are **different domains** yet they are interdependent.

d'Arcy

$x$ GETS CLOSER AND CLOSER TO ONE

When considering the limit of $f(x)$ as $x$ approaches 1, **$x$ gets closer and closer to one**, but we are not concerned about what happens when $x$ equals one.

---

For me $x$ is the unknown. It can stand for everything in the universe. **One** exemplifies unity, wholeness. So the terminology **$x$ gets closer and closer to one** suggests the interrelationship of all things and the direct connections between macro and micro environments. I choose to symbolize this concept by selecting one human who receives both physical and metaphysical messages from spheres as distant as the farthest star and galaxy. This person internalizes these messages through his brain and nervous system and allows these messages to permeate the billions of parts of his body. The upper sphere represents the universe. The lower sphere stands for the planet earth and the realm of the human body.

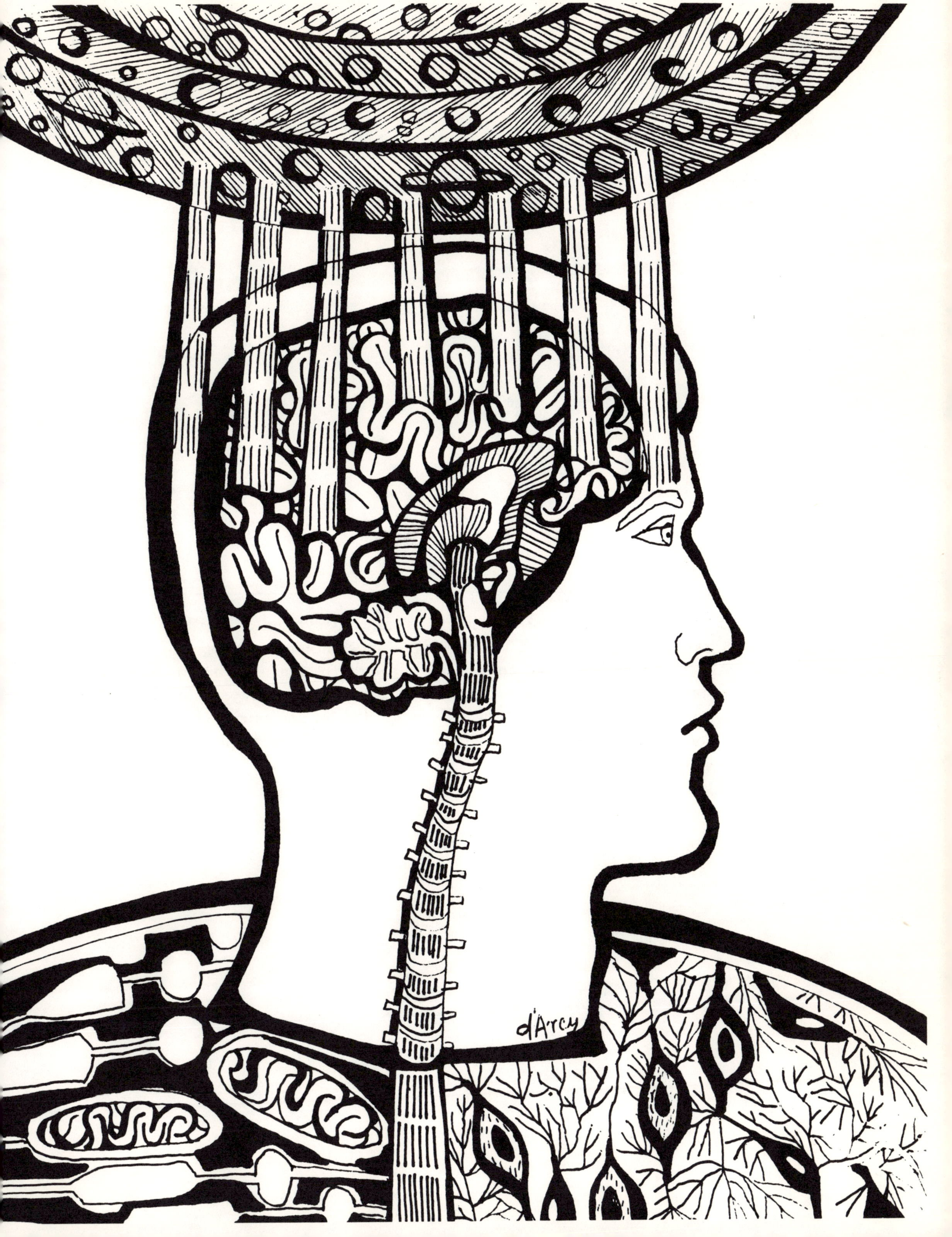
d'Arcy

MYSTERIOUS

I purposely don't write the algebraic expression in factored form because I like to make the result less obvious and more **mysterious**.

---

The river at the bottom of the drawing is an ancient symbol of spiritual cleansing and the everflowing source of life. The elephant, a most noble creature, long has been associated with wisdom and mystical powers. He is the earth-creature who has been given the privilege of transporting to all places of the world the great **mystery** of life. The palace on his back houses this celestial **mystery**. He is guided on his journey by the triangle in the heavens and reflected on his back which represents the **mysterious** aspect of the universe.

d'Arcy

## AN IRRATIONAL EXPONENT THAT MAKES SENSE

In algebra we can not define an irrational exponent; that is we can not define what is meant by $a^x$ when $x$ is an irrational number. To arrive at a definition of **an irrational exponent that makes sense** we use calculus by first considering the natural logarithmic function and then defining the natural exponential function as its inverse and then defining $a^x$ as $\exp(x \ln a)$.

---

The face below is of a person in a dream state. The tree, rooted in his unconscious, is the form of his vision. The three figures, dressed in exotic costumes perched on the branches of the tree, bring beautiful delights that please the senses and allow for a complete escape into a different reality. The wavy lines above the three figures of the dream-vision suggest an aura that crowns the scene with a sense of magic. This is indeed one **irrational exponent that makes sense**.

d'Arcy

THIS FUNCTION OBEYS

The natural logarithmic function is defined by

$$\ln x = \int_1^x \frac{1}{t}\,dt, \qquad x > 0.$$

**This function obeys** the properties of logarithms you learned in algebra.

---

The word **obey** brings to mind an act of servitude. One **obeys** an order whether or not one agrees with or even understands the command. Some ancient cultures were ruled by monarchs who had absolute power over all of their subjects. These people **obeyed** or suffered the consequences. On the other hand when an individual chooses **obedience** as a way of life, as in some religious orders, the same **obedience** can become a haven, a place of full trust, a place of peace, and a welcome relief from the often painful process of making decisions. The notion of **function** can be seen as a certain duty or performance by a given thing. I once wrote a book on the **functions** of art in which I selected ten from the endless number of **functions** that art serves in society: discovery, intensification, expression, record, communication, interpretation, reformation, enhancement, order, integration.

For this drawing, I chose the **function** of technology as the focus of my attention because it seems most suited to the idea of **obedience**. The "technology man," dressed in prison stripes and bowing low before his human master, becomes the 20th and 21st century slave who perfectly **obeys** all commands whether or not they are wise ones.

d'Arcy

INTERVALS

We determine the **intervals** on which the function $f$ is increasing by finding the values of $x$ for which $f'(x)$ is positive. To determine the **intervals** on which $f$ is decreasing we find the values of $x$ for which $f'(x)$ is negative.

---

**Intervals** are as basic to our existence as the breaths we take, the beats of our hearts, and in a larger context, the tides, the seasons, the movement of the planets. While **intervals** play a part in all of the arts, I especially associate them with music.

This drawing makes reference to music through the depiction of musicians performing on a variety of musical instruments from different periods of time and several cultures. We must imagine the **intervals** of sound they create through the notes of the flute, the plucked strings of the lute and harp, and the beat of the drum. We also are reminded of **intervals** by the architecture represented here. The style I have chosen, loosely based on traditional Japanese construction, uses line and space to produce an aesthetically pleasing set of **intervals**. Turning again to the Asian love of nature and attention to the seasons, we see that the drawing contains reference to one of the seasonal **intervals**, spring, with the presence of the old trees dressed in budding blossoms.

d'Arcy

BOTH SIDES EXIST

The definition of an irrational exponent is as follows: If $a$ is any positive number and $x$ is any real number, then

$$a^x = \exp(x \ln a).$$

In this equation **both sides exist**.

---

The two main shapes of the picture plane represent two neurons of the human brain that meet and make a vital connection which leads to all other human functions and communications. One neuron contains a female and the other a male. This symbolizes the vitality and strength that results in a meeting of opposites and supports the importance of the notion that **both sides exist**. The same concept applies to the workings of the cosmos. The circles that float around the central figures are marked with the Yin-Yang symbols of Taoist and Buddhist tradition which stand for the inherent structure of complementary opposites in all things wherein **both sides exist**.

d'Arcy

INTEGRATING A POWER AND PERFECTLY RIGOROUS

The formula for **integrating a power** is

$$\int u^n\,du = \begin{cases} \frac{u^{n+1}}{n+1} + C & \text{if } n \neq -1 \\ \ln|u| + C & \text{if } n = -1 \end{cases}.$$

Our proof of this formula was **perfectly rigorous**.

---

While stereotypes never tell the whole truth, I will give in to the temptation of generalization for the sake of art in this case. The words **power** and **perfectly rigorous** somehow bring to my mind a macho-male picture. This drawing is meant to evoke masculinity. The strength, the **rigorous** strenuous nature, and the aggressive action of men, that we can both fear and be attracted to at the same time, are pictured here. The climate caused by these opposing emotions can, for some of us, create an erotic environment.

d'Arcy

INFINITY

We use the symbols $+\infty$ (positive **infinity**) and $-\infty$ (negative **infinity**) to indicate the behavior of a function. Stating that the limit of $f(x)$ as $x$ approaches $a$ is $+\infty$ means that the function values increase without bound as $x$ gets closer and closer to $a$. Stating that the limit of $f(x)$ as $x$ approaches $a$ is $-\infty$ means that the function values decrease without bound as $x$ gets closer and closer to $a$.

---

The great and recurring wave of humanity surges into **infinite** space after having spent its allotted time on earth. The forces of the universe that give direction and energy to this new birth of the spirit are represented by the lines accompanying the wave. All is delivered into the loving hands of an angel who represents the beauty of the unknown, the **infinite**.

d'Arcy

THE LANGUAGE OF CALCULUS

The mathematical symbolism

$$\lim_{x \to a} f(x) = +\infty$$

is part of **the language of calculus**. It can be read as "the limit of $f(x)$ as $x$ approaches $a$ is positive infinity."

---

The lower figure stands for the realm of letters and words. The upper figure represents the realm of numbers and mathematics. The two combine to form a major foundation for civilization. The man is reading words from ancient and modern languages. He is also reading **the language of calculus** which is being spoken by the woman above. Words and numbers combine to form powerful modes of communication.

AMOUR
EST
B
G
A
N
Y
H
J
Z
D
F
AR

DIFFERENTIATION

The process of computing the derivative is called **differentiation**. Thus **differentiation** is the operation of deriving a function $f'$ from a function $f$.

---

The term **differentiation** sends me in two directions. First, I take its meaning relating to dissimilar, distinguishing, unusual, other. Here, I am reminded of my film-making journey into the remote estuaries of the Amazon River in Brazil where I encountered the most extraordinary abundance and variety of natural forms. It seemed to me that every plant or tree, insect or animal, I saw was a totally **different** one from all the rest. This seemingly endless set of **differences** was almost unbelievable until I thought of the fact that each human being on earth has his own finger prints, **different** from the billions of others. And when I turn to the arts, there is also an astounding **differentiation**. Take the octave in music. Seven notes that have been arranged in such **different** sequences and tempo that they form a compendium of melodies, sonatas, symphonics, operas, folk songs, jazz, rock, and other new age musical forms that stretch on and on. The same is true of the visual and literary arts, and in whole cultures where humans in **different** environments have created such **different forms** in their lifestyles, food, shelter, costumes, adornments, ritual, recreation, religion, and so on.

The second set of images I receive from the term **differentiation** is that having to do with the all important relationship between time and space. In the drawing here, I have used a mechanical reference, the **differential** gear, to express the concept that the smaller unit with the infant has its own power to turn the larger gear with its more lumbering rhythm of adult humans and animals. Both of these **differentiating** qualities are significant.

d'Arcy

INTEGRATION

The process of evaluating an indefinite integral or a definite integral is called **integration**.

---

The mandala, a circular form that appears in the art and philosophy of many cultures, was a central force in the structure of ancient Indian society. This mandala is traditionally intersected by a central cross which divides the circle into four parts. The rest of the mandala design, including thousands of intricate motifs, is a tremendously complex guide to the four aspects of life and to the **integration** of these separate entities into a whole.

My drawing uses the basic mandala form of circle and central cross but takes off on a somewhat whimsical human trip. I have placed the eye, ear, nose, and mouth in each of the four parts to represent the senses of sight, hearing, smell, and taste. It is my nope that the viewer of this drawing will come up with his own mental process of **integration** to form the whole individual of his choice.

d'Arcy

CONTINUITY

Probably the most important class of functions studied in calculus are **continuous** functions. We begin our study of such functions by considering **continuity** of a function at a number. We then discuss **continuity** of a function on an interval.

---

In the human population of this earth, there is an endless variety of sizes, shapes, colors, and structures as well as customs, beliefs, traditions, and social-political systems. For this drawing, I have selected the human heart, with its symbolism of love and compassion, to express the **continuity** that exists within all human beings.

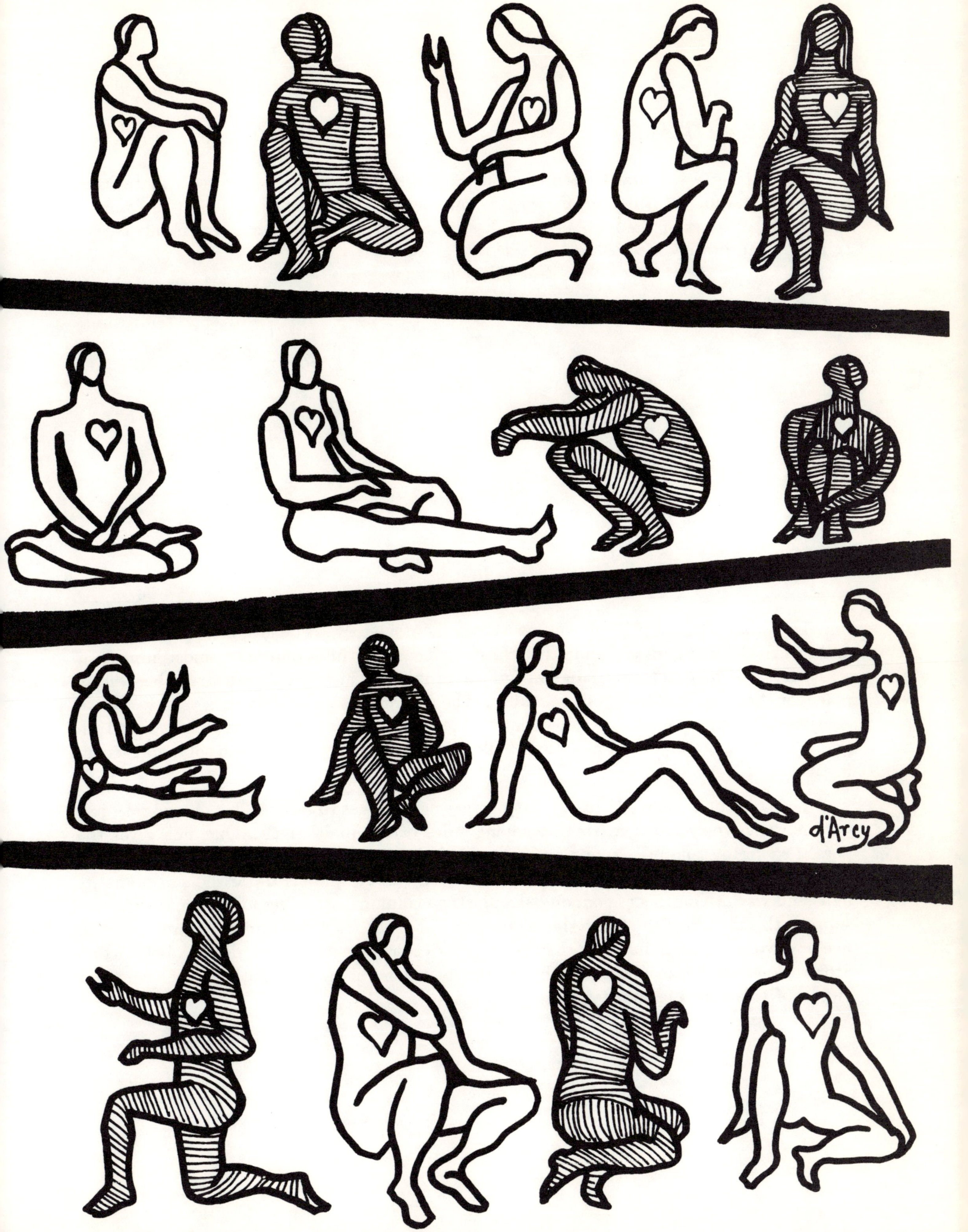
d'Arcy

TRIANGLE INEQUALITY

The **triangle inequality** is used to prove some of the limit theorems. It states that the absolute value of the sum of two real numbers is less than or equal to the sum of the absolute values of the numbers. With symbols, where $a$ and $b$ are real numbers, the **triangle inequality** is written

$$|a+b| \leq |a| + |b|$$

---

The triangular form has been used throughout history to denote strength, stability, and security. The Egyptians and other cultures of earliest times built their most important structures, the pyramids, in triangular form to provide for an eternal, indestructible resting place for their spiritual and political rulers. The pediment on ancient Greek temples which continued to appear ever since on various periods of western architecture, including 19th and early 20th century court and bank buildings, gave the impression of a solid and secure institution.

The concept of **triangle inequality** brings new images to mind. This drawing depicts two humans who inhabit a triangle having sides of unequal length. One side gives the person room to stand up, expand and grow to his fullest potential. The spirals in this part of the triangle suggest growth and progress. Even outside the **triangle** the sun pours its light and warmth on this region, and strong stone columns, standing for accomplishments, support this section of the **triangle**. The other end of the triangle requires its inhabitant to live and work eternally in a crouched position. Here we see all human potential thwarted. Storm clouds and rain allow very little light to reach this unfortunate person, who is a victim of **triangle inequality**.

d'Arcy

## DEGENERATE CONIC

The graph of the equation

$$Ax^2 + Cy^2 + Dx + Ey + F = 0$$

where $A$ and $C$ are not both zero, is either a **conic** or a **degenerate conic**. A **conic** is a curve of intersection of a plane with a right-circular cone. There are three types of curves that occur this way: the parabola, the ellipse, and the hyperbola. For instance, an ellipse is obtained as a **conic** if the cutting plane is parallel to no generator, in which case the cutting plane intersects each generator. The **degenerate** case of an ellipse, a point, is obtained as a **conic** if the cutting plane contains the vertex of the cone but does not contain a generator.

---

In socio-culture terms a **degenerate** person is one whose lifestyle does not conform to the social and moral standards of his community. He is seen to be one who has fallen from all accepted standards of behavior and is viewed as a sick and troubled person.

This drawing depicts the horrific plunge of a **degenerate** man into the hungry flames of hell. The vehicle conveying him to this fiery inferno is in the form of a cone. This is why we call him a **degenerate conic**.

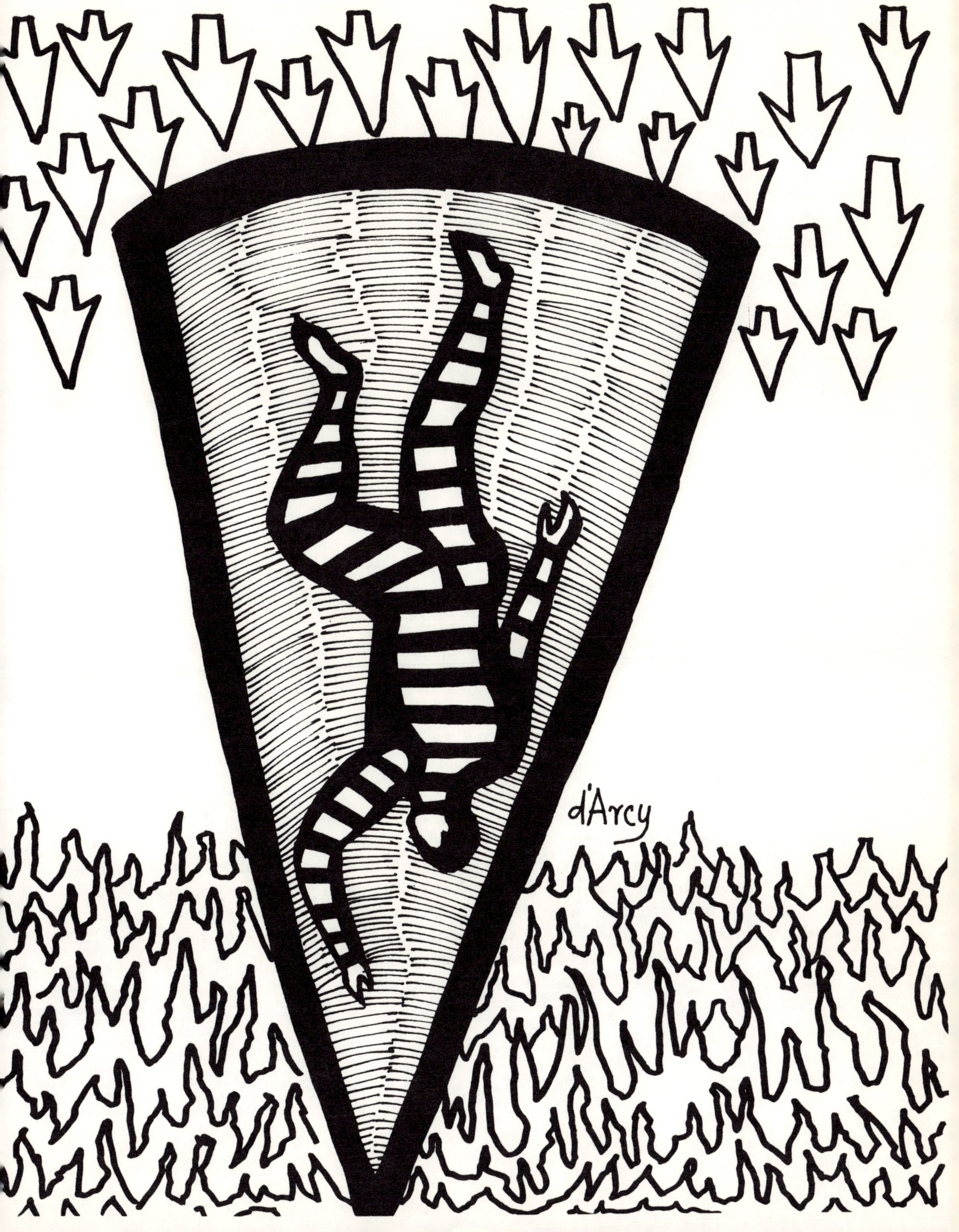
d'Arcy

TRANSCENDENTAL

Functions that are not algebraic are called **transcendental**. Examples of **transcendental** functions are the trigonometric, exponential, and logarithmic functions. The number $e$ is a **transcendental** number; that is, it cannot be expressed as the root of any polynomial with integer coefficients.

---

I feel that **transcendence** is at the heart of every human spirit. The concepts of rising above, going beyond, and surmounting the sometimes ugly realities of life are aspirations we all share. As evidenced in the religions, philosophies, myths, and legends of every culture since the beginning of humanity, we have something deep within our psyches that makes us sense that there is another plane of being, a **transcendental** level of consciousness that we can reach only through some form of dedicated search, such as found in meditation, devotion, or prayer.

The scene of this drawing is on the top of a mountain. The steep cliffs have paths and staircases to aid the **transcendental** climber in the attainment of his goal. But we must note that some of these stairs and pathways lead nowhere, suggesting that the process of finding one's way in the metaphysical landscape is not an easy one. But with patience and persistence, the spiritual traveller can make his way to the top and full **transcendence**. The rewards take the form of three symbols: the human tree representing man's oneness with nature; the person touching sun and moon that suggests cosmic harmony; and the woman in the hand which depicts human creativity.

d'Arcy

VALUES

If $f$ is a function, then $f(x)$ denotes the **value** of $f$ at the number $x$; that is, $f(x)$ is the number in the range of the function $f$ corresponding to the number $x$ in the domain. New functions are formed from given functions by adding, subtracting, multiplying, and dividing function **values**. Accordingly, these new functions are known as the sum, difference, product, and quotient of the original functions.

---

The application of the word **values** to many different disciplines and aspects of society is indeed broad. This term occurs in mathematics and the other sciences, the visual and musical arts, psychology, sociology, and commerce, as well as in all of the humanities including philosophy and, of course, religion. In particular, in the visual arts, **value** can refer to the relative lightness or darkness of a color or the effect produced in a work of art by the use of light and shade. In music, **value** can pertain to the duration of a note, tone, or rest.

But the concept of **values** goes to the heart of all civilizations in its widest meaning. Here the word stands for the social and ethical principles, goals, or standards held by an individual, group, or an entire society.

In this drawing the symbol for **values** is the diamond shape in the heart of the woman's body. The other figures represented here depict her life experiences which contribute to the forming of her **values**. The bands, criss-crossing her body and extending outward and upward, stand for those mysterious forces of the universe that also help to nourish her **value** system.

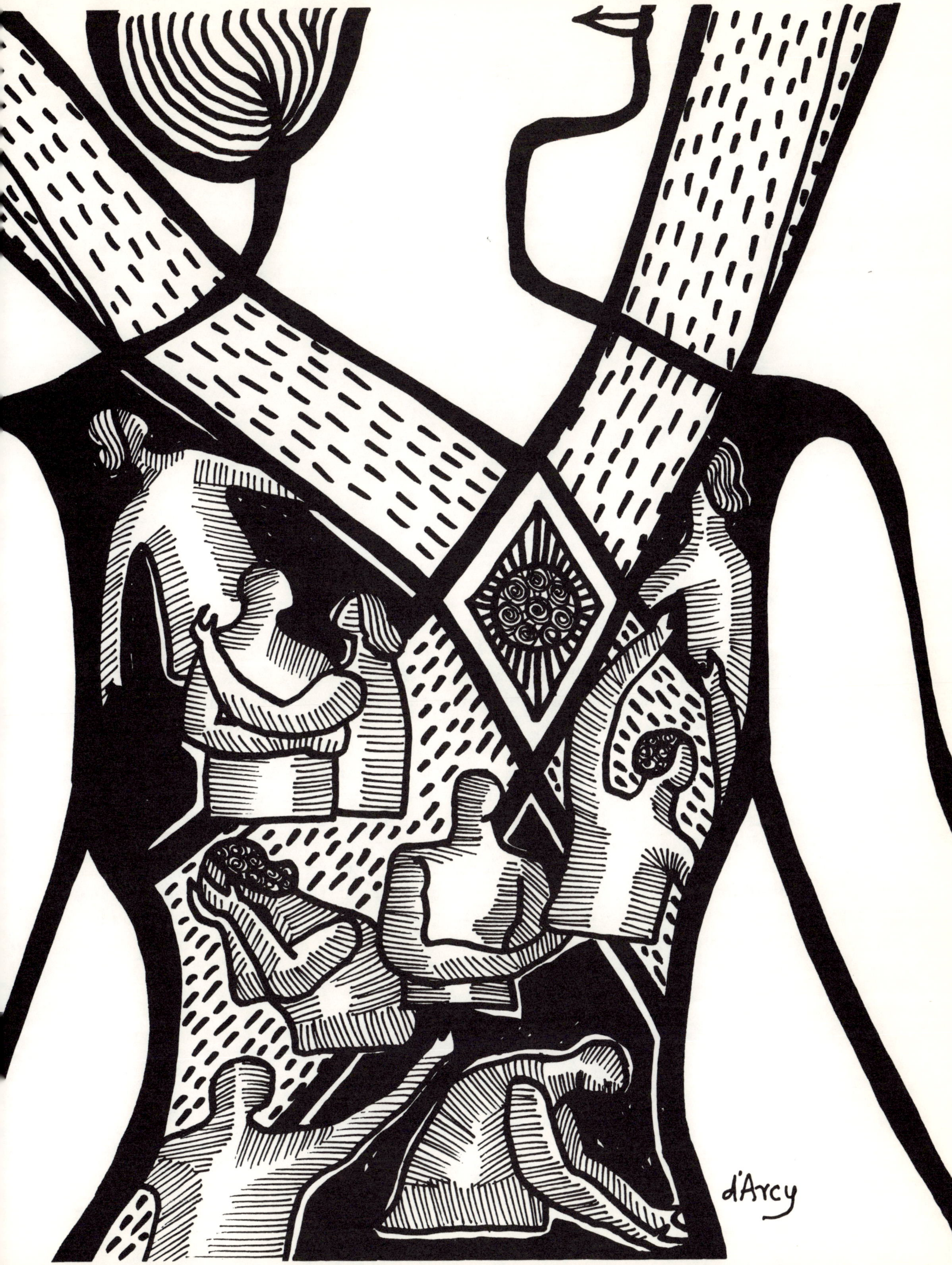
d'Arcy

BOUNDARY CONDITIONS

Often in applications involving antidifferentiation, it is desired to find a particular antiderivative that satisfies certain **conditions**. If the **conditions** occur at one point they are called initial **conditions**. If they occur at more than one point they are called **boundary conditions**.

---

The notion of **boundary conditions** leads me down many paths. First I want to think about **boundaries**. These mark the lines between one thing and another. They delineate regions. They designate beginnings and endings. They tend to enclose, confine, and restrict, but they also can organize and protect. There are social-political **boundaries**, ethical and moral **boundaries**, physical and metaphysical **boundaries** among others. When the word **conditions** is added to make the term **boundary conditions**, I am lead to the images of this drawing. There are three **boundary** structures which require the individuals and groups of people within them to live in grim **boundary conditions**. These may be national, legal, or racial boundaries that can provide safe environments for their inhabitants, but one suspects that such narrow confines frustrate, and finally suffocate, those who are forced to dwell here. The one thing lacking in these structures is freedom. The three birds flying above symbolize freedom, that aspect of human rights most sought by those who must exist in strict **boundary conditions**.

'Arcy

## ROUND-OFF ERROR

To consider the accuracy of the approximation of a definite integral by the trapezoidal rule, we refer to two kinds of errors that are introduced. One is the error due to the approximation of the graph of the function by segments of straight lines. The term truncation error refers to this kind of error. The other kind of error, which is unavoidable, is called the **round-off error**. It arises because numbers having a finite number of digits are used for approximations.

---

To err is human. All western cultures know this. Eastern cultures take this idea one step further. Asian artists incorporate this concept in their work. For example, traditional Korean potters, whose ceramic works are considered to be among the greatest in the world, have perfected their skills in this ancient art to a point where they can produce a technically faultless object every time. But because such an artist believes that no mortal being has the right to presume that he can be perfect in any aspect of life, as this quality belongs only to God, he consciously makes one **error** in every piece, be it in the glaze or shape of the object, so that he remains a humble, respectful human being.

To illustrate the phrase **round-off error**, I have invented a small fantasy. A great river has **rounded off** the top of a huge cliff. By some navigational **error** these ships of yore have not foreseen this enormous drop and must now face the unfortunate consequences.

d'Arcy

ORDERED PAIR

Any two real numbers form a **pair**, and when the **order** of appearance of the number is significant, we call it an **ordered pair**. If $x$ is the first real number and $y$ is the second, this **ordered pair** is denoted by $(x, y)$. Observe that the **ordered pair** $(3, 7)$ is different from the **ordered pair** $(7, 3)$.

---

This drawing presents three **pairs** of humans. Each is an **ordered pair** as can be determined by the matching suits they are wearing. **Order**, having to do with the arrangement of things, sets out a fixed plan to designate position, priority, system, or any other categorization. In this case, **pair** number one combines two women, **pair** number two is made up of a man and a woman, and **pair** number three consists of two men. They are all **ordered pairs**.

1
3
d'Arcy
2

FUNDAMENTAL IDENTITY

The relationship between the sine and cosine functions given by the equation

$$\sin^2 t + \cos^2 t = 1$$

is a **fundamental identity**. It is an **identity** because it is true for all real values of $t$. It is **fundamental** because it is used to obtain other **identities**.

---

**Identity** and individuality are closely related. The artist often establishes his **identity** through his work. However, the projected image of self through art or any other expressive medium can be far from existing fact. This is where the inventiveness of the mind allows for the creation of totally new fantasy **identities** that please the wearer more than his own skin. Then there is the human potential for the **identification** of oneself with someone else. This is where role-models, heroes, and even Andy Warhol's super stars, famous for only fifteen minutes, can shape the projected **identities** of a whole generation.

This drawing shows two people projecting **fundamental identities** quite different from themselves.

dArcy

# RIGHT-HANDED SYSTEM AND LEFT-HANDED SYSTEM

To represent the set of all ordered triples of real numbers in a geometric three-dimensional space we consider the directed distances of a point from three mutually perpendicular planes. The planes are formed by three mutually perpendicular lines that intersect at a point called the origin. These lines are designated as the $x$ axis, the $y$ axis, and the $z$ axis. Usually the $x$ axis and $y$ axis are taken in a horizontal plane, and the $z$ axis is vertical. The coordinate system is called either **right-handed** or **left-handed**. We have a **right-handed system** if when the right hand is placed so the thumb is pointed in the positive direction of the $x$ axis and the index finger is pointed in the positive direction of the $y$ axis, then the middle finger is pointed in the positive direction of the $z$ axis. If the middle finger is pointed in the negative direction of the $z$ axis, then we have a **left-handed system**.

---

The **right hand** contains a **right-handed** person who is framed by arrows rotating in a clockwise direction. The fabric of her garment contains clockwise spirals. The **left hand** contains a **left-handed** person surrounded by counter-clockwise arrows and wearing counter-clockwise spirals. The bands emanating from the fingers extend far into the **left and right-handed** universe.

d'Arcy

EMPTY SET

A **set** is a collection of objects, and the objects in the **set** are called elements. The **set** that contains no elements is called the **empty set**. For example, the **set** of all real number solutions to the equation

$$x^2 + x + 1 = 0$$

is the **empty set**.

---

The word **empty** brings different images and symbols to different cultures. In western civilization **empty** tends to suggest void of contents, nothingness, uselessness, and worthlessness. Almost the reverse is true in many eastern philosophies, where **emptiness** and nothingness are considered to be states of being of the greatest purity and spiritual attainment. In Hindu, Buddhist, and other Asian teachings, one is encouraged to strive toward this holy state of enlightenment even if it takes a lifetime of concentration and meditation. **Emptiness**, nothingness, and nirvana are one.

The fantasy of this drawing is the following: The devoted son of a great maharajah wanted to surprise his father on his sixtieth birthday with a magnificent gift. He commissioned hundreds of architects, builders, and craftsmen to construct the most intricate and ornate palace ever known to serve as his father's new home. After living in this abundance of decoration for one year, the father, on the brink of suffocation, ordered that a space be cleared in the middle of the palace where he could meditate in peace. This **empty set** became the father's favorite place of refuge.

d'Arcy

## SYMMETRY

To sketch the graph of an equation it is often helpful to consider properties of **symmetry** of the graph. Two points $P$ and $Q$ are said to be **symmetric** with respect to a line if and only if the line is the perpendicular bisector of the line segment $PQ$. Two points $P$ and $Q$ are said to be **symmetric** with respect to a third point if and only if the third point is the midpoint of the line segment $PQ$.

---

**Symmetry** involves a correspondence of opposite parts. While this can be seen to be related to the ancient Asian concepts of the complementary aspects of opposites, it might be noted that it was in western art traditions that **symmetry** flourished. From early Greek architecture and sculpture through Renaissance painting and other visual arts leading all the way to the early part of the 20th century, it was generally considered that the highest form of balance and beauty resulted from a **symmetrical** composition. Many formulas and sets of measurements were proposed by art historians and other scholars to prove the superiority of **symmetry** in art. All this occurred while, of course, eastern cultures ignored these pronouncements, and did quite well with their own **asymmetric** arrangements of the aesthetic elements.

This drawing illustrates the concept of **symmetry**.

d'Arcy

ONE-TO-ONE

A function is said to be **one-to-one** if every number in its range corresponds to exactly one number in its domain. We know that a vertical line can intersect the graph of a function in only one point. For a **one-to-one** function, it is also true that a horizontal line can intersect the graph in at most one point.

---

Communication has become one of the most important elements of society. If our civilization is to survive and develop and if the human species is to continue at all, our ability to communicate must be protected, preserved and even improved. The most direct form of communication is **one-to-one**. As simple as it may seem for communication to take place between two individuals, it involves extremely complex processes. When people understand each other in depth it is a rare phenomenon.

This drawing shows two persons from different cultural backgrounds, with different genes, as symbolized by the patterns on their bodies and different views of the universe taking the first step in their ongoing effort to communicate **one-to-one**.

d'Arcy

## COMPONENT OF THE FORCE

If a constant **force** of $F$ pounds moves an object a distance $d$ feet along a straight line and the force is acting in the direction of motion, then if $W$ is the number of foot-pounds in the work done by the force, $W = Fd$. Suppose however, that the constant force is not directed along the line of motion. In this case the physicist defines the work done as the product of the **component of the force** along the line of motion times the displacement.

---

A **component**, one of the parts of a whole, has two vital roles to play. In itself, it acts as an independent entity, valid within its own dimensions. Combined with other elements, it gains new strength and serves as a team member.

This drawing concentrates on the **component of the force**. The **force**, crashing into an opposing barrier, accumulates its power from its many **components**, including the one **component** receiving our special attention.

d'Arcy

THE CHAIN RULE

To compute the derivative of a composite function we use one of the important theorems in calculus called **the chain rule**. The theorem states that if the function $g$ is differentiable at $x$ and the function $f$ is differentiable at $g(x)$, then the composite function $f \circ g$ is differentiable at $x$, and

$$(f \circ g)'(x) = f'(g(x))g'(x)$$

---

Strong **chains** border the two levels of activity over which the **chains rule**. The top level shows humans bending over in hard physical labor in an agricultural field. The lower level pictures beasts of burden, oxen pulling carts of huge rocks.

d'Arcy

MOMENT OF MASS

Consider a horizontal rod, of negligible weight and thickness, placed on the $x$ axis. On the rod is a system of $n$ particles where the $i$th particle is at a directed distance $x_i$ meters from the origin and its **mass** is $m_i$ kilograms. We define the number of kilogram-meters in the **moment of mass** of the $i$th particle with respect to the origin as $m_i x_i$. The **moment of mass** for the system is defined as the sum of the **moments of mass** of all the particles.

---

The two people in the drawing, while passing an uneventful day in this desert terrain, suddenly discover a huge **mass** hovering overhead. They wonder if it could be an extraterrestrial form, a meteorite, or even a space ship. But before they can further study it, the thing disappears. It was only a **moment of mass**.

d'Arcy

GROWTH AND DECAY

The laws of **growth and decay** arise when the rate of change of the amount of a quantity with respect to time is proportional to the amount of the quantity present at a given time.

---

In life there is **growth and decay**. I see **growth** as an audacious defiance of gravity. A tiny plant, which challenges the very power of the whole planet by insisting on its own movement up and away from the earth, is one of the great miracles. **Decay** is the opposite truth, and in the physical world it is a process just as important as **growth** because it helps nature make space and clear the way for new birth.

The drawing depicts three plant forms that have grown to maturity and have begun to turn back toward the earth in **decay**. At one point in these plants' development, certain forces signal the end of the growth stage. These forces are represented by the dark panels. The remains of the plants' **decay** is shown in the horizontal forms on the ground. The parallel human **growth and decay** is symbolized by the babies and young children at the base of the plants followed by people of different cultures in their various stages of development until they reach old age and physical **decay**.

d'Arcy

HALF-LIFE

In problems involving the law of natural decay, the **half-life** of a substance is the time for half of it to decay.

---

This drawing is a bit of whimsy which, as you can see, is the **half-life** of a man as well as the sun.

d'Arcy

RADIOACTIVE SUBSTANCE

Suppose the rate of decay of a **radioactive substance** is proportional to the amount present. This is an example of exponential decay. If the half life of the substance is known, we can determine how long it will take for any given amount of the substance to disappear by solving the differential equation

$$\frac{dA}{dt} = k\,dt \qquad k < 0$$

with the appropriate boundary conditions.

---

When I think of a **radioactive substance** which gives off radiant energy in particles or rays by the disintegration of the atomic nuclei, I always picture the mushroom cloud produced by an atomic explosion. Instead of taking you with me into my personal nightmare and underlying fear of the destruction of the entire planet, I have decided to dwell in this drawing on the form of the mushroom and its magical mystical powers throughout the ages in all parts of the world.

The shamans in the Amazonian jungles of Brazil and many other indigenous peoples in Latin American countries continue the traditions of ancient cultures in Asia, Africa, Europe, and North America in their use of mushrooms to produce visions and other states of ecstasy within the context of religious experience. The look of a mushroom, the way it grows, its spermal fragrance, and its positive and negative powers, all add to its mystery and, in my view, its role as a **radioactive substance**.

d'Arcy

FAMILY OF SOLUTIONS

The complete **solution** of the differential equation

$$dy = f(x)\,dx$$

is given by

$$y = F(x) + C$$

where $F$ is an antiderivative of $f$ and $C$ is an arbitrary constant. This equation represents a **family of solutions**. Because there is one arbitrary constant, it is called a one-parameter **family**.

---

The word **family** brings many thoughts to mind. First, there is the understanding that a **family** represents relationships between different objects or individuals. Then we can view the human **family** as the fundamental form in a civilization. **Families** include likenesses, differences, cohesive strength, tensions, favoritism, cooperation, jealousy, competitiveness, sharing, love, and hate, among other things. According to the Judaic-Christian view of the first **family**, as recorded in the Bible, Adam and Eve's **family** life was far from serene with disputes between the brothers Cain and Abel leading all the way to murder.

This drawing depicts a **family** in one of its more peaceful moments, contemplating possible **solutions** to a problem at hand. The boxes above their heads contain the **solutions** each proposes. While some **solutions** are similar to others, each has the stamp of an individual human brain, always unique in its structure and output.

d'Arcy

ABSOLUTE EXTREME

An **absolute extreme** function value on an interval is either the greatest function value (**absolute** maximum value) or the smallest function value (**absolute** minimum value) on the interval.

---

One of the teachings of the Buddha was to avoid the **extremes**. This reflected his general view of the complementary aspects of opposites. The two women in this drawing have not learned this lesson. They have taken life positions at opposite extremes. The curve between them, representing the varying degrees of compromise and agreement, is a path they refuse to take. The sphere in the sky above the curve represents the **absolute**. Each woman is rigid and keeps her gaze on the **absolute extreme**.

dArcy

POLAR COORDINATES

The position of a point in the plane can be located by a system of **coordinates**. In particular, the cartesian **coordinates** of a point are numbers that are the directed distances of the point from two fixed lines. The **polar coordinates** of a point consist of a directed distance and the measure of an angle, which is taken relative to a fixed point, called the pole, and a fixed ray, called the **polar** axis.

---

Before I discuss this drawing, I want you to be aware of the fact that I am aware of the fact that **polar** bears do not inhabit the **polar** regions of our planet. I think the same goes for penguins.

Thank God for "artists' license." So here we have these northern creatures straddling ice banks and looking cold. All this just to make a visual statement about **polar coordinates**.

N
W
E
S
d'Arcy

## A DENT OR A LOOP

The graph of the equation

$$r = a + b\cos\theta$$

where $a$ and $b$ are positive numbers is a limaçon. The limaçon may have **a dent or a loop**. If $0 < a/b < 1$, it has a **loop**, and if $1 < a/b < 2$, it has a **dent**.

---

The landscape in this drawing is made up of a series of **dents**, while the sky is hung with **loops**. Each contains a face. For me there is a stillness here, a place of quiet and peace. It could be that these are souls awaiting their final destiny. This might be a purgatory garden.

d'Arcy

SPIRAL

**Spirals** occur frequently as curves having equations in polar coordinates. The graphs of the equations

$$r = 2\theta, \qquad r = e^{\theta}, \qquad r = \frac{3}{\theta}$$

are **spirals**. The graph of the first equation is a **spiral** of Archimedes, the graph of the second is a logarithmic **spiral**, and the graph of the third is a reciprocal **spiral**.

---

The **spiral** is one of those fundamental forms in both the physical and metaphysical worlds. All cultures have employed the **spiral** in the visual arts. It is the most basic symbol for growth and development. The unfolding lines of the **spiral** convey the message of evolution. Its continuum suggests an eternal progression.

This drawing lets the **spiral** form the basis for the growth and refinement of the universe. First we see the stars and planets, next we concentrate on life on earth with flora and fauna taking shape, and finally there is the appearance of humanity.

d'Arcy

INTERSECTS ITSELF

The graph of the polar equation

$$r = \sin\left(\frac{3}{2}\theta\right)$$

**intersects itself** at the points $\left(\frac{1}{2}\sqrt{2},\ \frac{1}{6}(2n+1)\pi\right)$, where $n$ is $0, 1, 2, 3, 4, 5$.

---

To **intersect** or to divide something into two parts is one action; the notion of something **intersecting itself** is more complicated and thus, more intriguing. The man in this drawing is **intersecting himself**. Why? We will never know. But, this strange ritual is not much more baffling than the thousands, perhaps millions, of weird things human beings have done since the beginning of history. Some of these peculiar rituals are carried on in the privacy of one's life and are kept as dark secrets. But others of these seemingly senseless acts become traditions and fashions, shared by small or large groups of people and even spreading to whole cultures over vast periods of time. A few of these customs can be noted here: piercing ears, noses, tongues, or a number of other body parts to insert jewelry; binding feet or heads to change their shape and make them more attractive to local tastes; wearing high-heel shoes; performing cosmetic surgery on eyes, chins, noses, ears, breasts, and so on, to make these features conform to current ideal standards.

The man in the drawing is tattooed all over his body. This custom in Thailand is instituted to protect those males who work in the rivers and canals from poisonous water snakes. In other cultures, it is done for aesthetic enhancement and for symbolic purposes.

MOTHER
d'Arcy

POSITIVE AND NEGATIVE

Because the domain of the natural logarithmic function is the set of positive numbers when we write $\ln x$, $x$ must be positive, but when we use absolute value bars and write $\ln |x|$, $x$ may be both **positive and negative**.

---

**Positive and negative** symbols in the physical world can represent opposing forces that complement each other to become energy sources. In reference to the human psyche, we can find different meanings for these two words. We associate **positive** with such terms as constructive, optimistic, uplifting, and so on, while **negative** often has the connotation of denial, refusal, rejection, and disapproval.

This drawing illustrates the **positive** with images of a mother and child. This mother who happily gives of herself to her offspring and who is rewarded by the miraculous growth and development of her infant, could be said to have a **positive** outlook on life. The large face below, reflecting on the tragic events that have taken place either in or around this life, depict a **negative** attitude.

d'Arcy

PERFECT COMPETITION

**Perfect competition** provides an application of calculus in economics. When a company is operating under **perfect competition**, there are many small firms; so any one firm cannot affect price by increasing production. Therefore, under **perfect competition** the price of a commodity is constant, and the company can sell as much as it wishes to sell at this constant price.

---

**Competition** is one of those natural phenomena that has never appealed to me. I do not feel attracted to or interested in contests, matches, or any other form of physical or social rivalry. And yet I recognize its place as a stimulant in society. Perhaps the one kind of **competition** that I can understand is that which one enters with oneself. This I consider to be the **perfect competition**.

Let's pretend that all the contestants in the running **competition** of this drawing are, in fact, different aspects of one person's personality. Competing here might be this individual's goodness, playfulness, studiousness, greed, selfishness, and so on. As the first three components seem to be winning, I judge this to be a **perfect competition**.

d'Arcy

## FOCUS

A parabola is the set of points in a plane equidistant from a fixed point and a fixed line. The fixed point is called the **focus** and the fixed line is called the directrix.

An interesting property of parabolas has applications to the construction of searchlights, automobile headlights, and telescopes. For a parabolic mirror in a searchlight, rays of light from a source at the **focus** are reflected along a line parallel to the axis. A similar principle involves parabolic reflectors in automobile headlights. For a parabolic mirror in a reflecting telescope, a converse situation occurs, where rays of light from an object in the sky, which go toward the mirror and are parallel to the axis, are reflected by the mirror and go through the **focus**.

---

The point where rays of light, heat, or sound come together is that precious meeting place called **focus**. In life, the physical ability to **focus** makes use of the eyes. The symbolic capacity to **focus** involves the brain and includes the ability to concentrate one's attention so as to observe and perceive a clear image or to form a precise idea.

This drawing considers the importance of the human eye and its recent extension, the camera, in the process of **focussing** as it relates to the arts. Of course, the visual arts, encompassing the ancient realms of architecture, crafts, painting, and sculpture as well as the more modern art fields such as photography, graphics, and design, all depend on the human eye. Perhaps the most powerful new human art tool is the motion picture camera. This superb invention and the whole field of art we call cinema provides humanity with a way to **focus** carefully on the human condition in different cultures and to disseminate this documented information to the widest public throughout the world.

d'Arcy

CIRCULATION

If $F$ is the velocity field of a fluid, the dot product $F \cdot T$ is the tangential component of $F$ and the line integral

$$\oint_C F \cdot T \, ds$$

is called the **circulation** of $F$ around the closed curve $C$. In an intuitive sense, we can think of the **circulation** as being the sum of the tangential components of $F$ around $C$.

---

Our lives depend on **circulation**. This drawing selects the **circulation** of blood in the body as one of the most vital forces within us. There are other physical forms of **circulation** that keep us going, such as air and nerve impulses. I feel that these patterns reflect the movement of the universe. As the stars and planets move in complete circuits, so do the intricate highways **circulate** within our physical homeland.

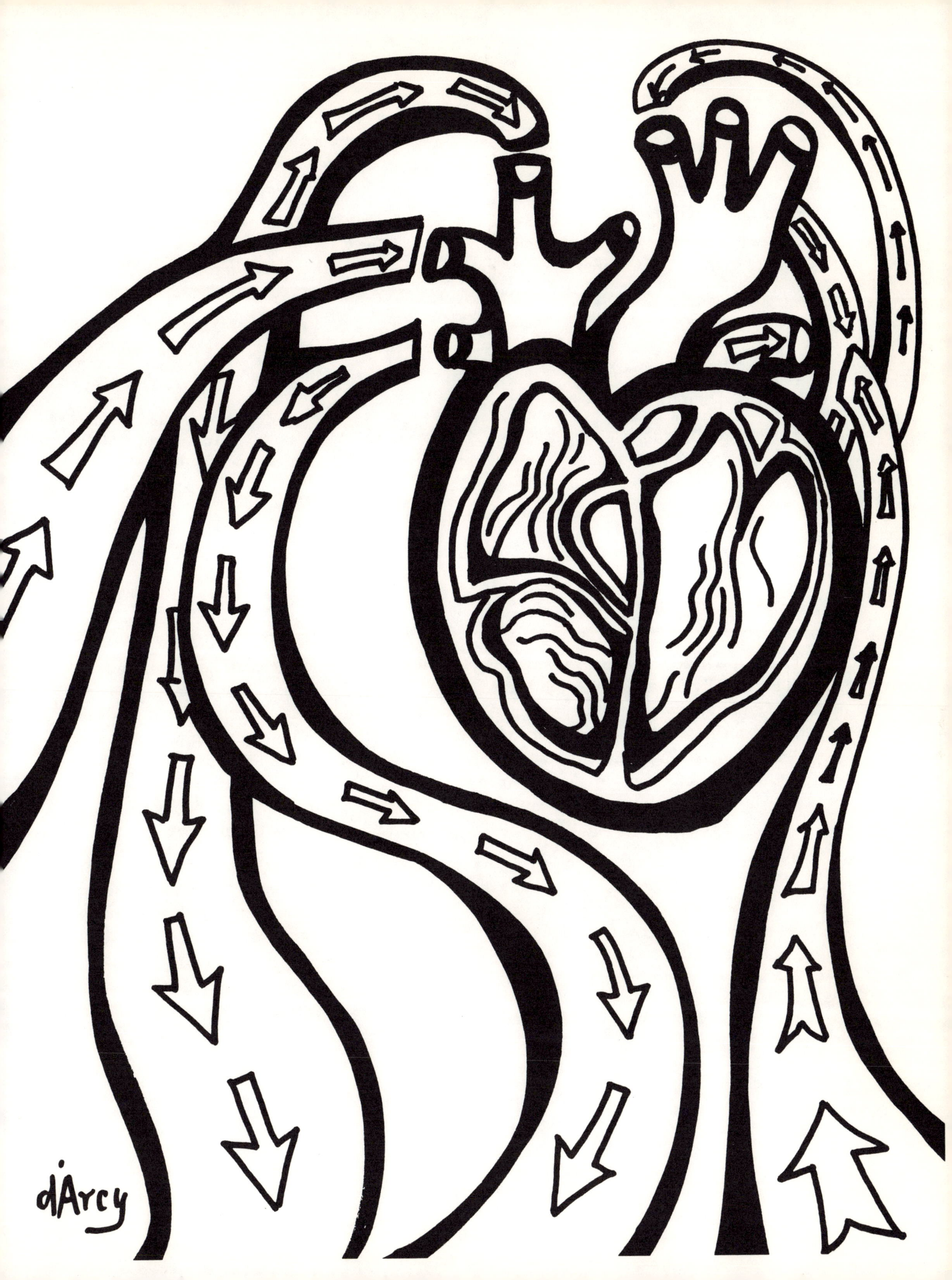
dArcy

FLUX

Suppose a fluid of constant density in a region $R$, having curve $C$ as its boundary, flows through $R$ in the counterclockwise direction along $C$. The flow of the velocity field $F$ across $C$ is the rate at which the fluid crosses $C$ in a direction perpendicular to $C$. This flow can be expressed as the line integral

$$\oint_C F(x, y) \cdot N(s)\, ds$$

which is called the **flux** of $F$ across $C$.

The word **flux** normally means flow; however, the terminology **flux** is applied to vector fields in general, not just to those associated with the velocity of a fluid.

---

The word **flux** has two meanings that carry powerful messages. The first refers to flow. This usually pertains to a liquid state as in the flow of the ocean tide or the river with its never ending, always flowing, water. This aspect of nature has become the symbol of spiritual cleansing and purification for many cultures. In eastern lands, such as India, the river serves as a shrine, where the body and soul can be washed and renewed. In Moslem cultures, water is of the utmost importance as a means of cleansing oneself five times daily before each of the five required prayers. Christianity, from its earliest times had considered water to be of the greatest significance and is used in all baptismal ceremonies.

The second meaning of the word **flux** is continual change. Einstein has said that the only constant is change. The power of change in society has been taken as a central symbol in contemporary western cultures. In a world often bound by tradition, many western countries have opened themselves to the new, to the unknown, to the young, and to continuous change. The result of this culture in **flux** has been the dynamic evolution of a new society where invention, innovation, and creativity thrive.

This drawing of a river illustrates the two meanings of **flux**: flow and change.

dArcy

CONSERVATION OF ENERGY

The sum of the potential and kinetic energies of a particle is constant along a curve; that is, the total energy of the particle remains unchanged during motion. This fact is a major concept in physics called the "law of **conservation of energy**."

---

The horse, a magnificent animal, has been portrayed in the arts throughout the ages as a symbol of unbounded **energy**. Even today, we measure the strength and **energy** of our machines in units of "horse-power." The two horses in this drawing are racing toward one another in unbridled **energy**. But they are contained within the barriers of dark lines that hold their **energy** in check, demonstrating the **conservation of energy**.

d'Arcy

ROOT TEST

To determine whether a given infinite series is absolutely convergent, both the ratio test and the **root test** are used. The ratio test is usually easier to apply, especially if the terms of the series contain factorials. However, if the terms of the series contain powers, it may be advantageous to use the **root test**.

---

In cultural terms, the notion of **roots** has taken on significant meaning, especially within the context of our own North American society. Just as the **roots** of a tree give strength and direction to that part of the tree which is visible above ground, the traditions and values of the individuals who people a land feed the general culture of that land. Diverse cultures have joined together to produce what we call the United States of America. We have been enriched, first by the native American heritage and then by European, Latin American, Middle Eastern, African, and Asian cultures. But just as a tree is only as healthy as its **roots**, a nation must continue to nourish the **roots** of its culture to remain strong.

This drawing shows a **root test** in progress. These three people are underground and they are examining the **roots** of the trees above. Happily, all is well as these **roots** receive a perfect bill of health.

d'Arcy

CLOSED BALL

A **closed ball** in two-dimensional space is the set of all points interior to and on the circumference of a circle. A **closed ball** in three-dimensional space is the set of all points interior to and on the surface of a sphere.

---

When you look at this drawing you may be tempted to join the festive activities at this fancy dress **ball**. So allow me to caution you to control your desires and let me quickly explain that this is a **closed ball**. Please note the securely fastened gate and heavy iron fencing surrounding the dancers. There is one more thing that I should point out to keep you from entertaining any notions of crashing this party: this is not only a **closed ball**; it is a **closed ball** within a **closed ball**.

d'Arcy

INDEPENDENT OF THE PATH

The value of a line integral is determined by the integrand and a curve between two points $P$ and $Q$. Under certain conditions, however, the value of a line integral depends only on the integrand and the points $P$ and $Q$ and not on the **path** from $P$ to $Q$. Such a line integral is said to be **independent of the path**.

---

Depicted in this drawing is the **path** of life that most people take. It is a perfectly good **path**, tried and true; it sets up the guidelines for a safe and generally well-balanced existence. Those who take this **path** like the company and approval of others. They do not want to take risks. They fear the criticism or ridicule of others if they veer off this beaten **path**. They are shown here wearing no clothes, which is meant to symbolize their reluctance to choose anything that might single them out as different from the others.

Then there is the rebel who makes his way in life **independent of the path** established by his forefathers and used by his peers. He could be an artist or scientist or philosopher or anyone else who sees another way and takes it.

There are probably both kinds of persons in all of us in different degrees. Again, much of this depends on when and where we were born.

d'Arcy

HIGHER ORDER

If the function $f$ is differentiable, then its derivative $f'$ is sometimes called the first derivative of $f$, or the first-order derivative of $f$. If the function $f'$ is differentiable, then the derivative of $f'$ is called the second derivative of $f$, denoted by $f''$. The derivative of $f''$, if it exists, is called the third derivative of $f$, denoted by $f'''$, and so on. These are derivatives of **higher order**.

---

In every culture since the beginning of human history, man's intuition or his experience of revelation, has provided a picture of a form of existence that transcends the everyday life of men and women on earth. This **higher order** can consist of gods and goddesses ruling from mountain tops and undersea dominions, or it might conceive of scholarly saints strolling through splendid gardens.

This drawing depicts a structuring of a **higher order** that architecturally organizes levels of spiritual attainment. It starts with the mortals below who are believers and striving to attain spiritual understanding. On the next level are those who, through devotion and prayer, are about to enter the long climb ahead. Then come the three levels of angels as well as the two guardian angels above, who are next to the very highest state. And finally at the top, seated in a carriage on the wheels of the universe, is the personification of one who has attained full knowledge and compassion in this **higher order**.

d'Arcy

# INDEX

**THE CALCULUS VIRGIN** is an unusual book that utilizes fantasy, humor, and highlights of the spirit "to celebrate the magic of human creativity in all its forms."

**d'ARCY HAYMAN** is an artist, a poet, a former faculty member of the University of California at Los Angeles and Columbia University, and from 1960 to 1980 head of the International Arts Program for the United Nations in Paris.

**LOUIS LEITHOLD** is the author of "The Calculus With Analytic Geometry," the six editions of which have been among the leading calculus texts used in universities, colleges, and secondary school advanced placement programs for more than twenty years. Writing this and other books, lecturing, and teaching have comprised his professional life.

THE CALCULUS VIRGIN resulted from one monumental day in the lives of these two friends. On this special day, Dr. Hayman requested permission to attend as an observer one of Professor Leithold's seminars on the teaching of calculus. He did not encourage her presence because he could not understand what she, having no mathematical background, would gain from attending. Later, however, after she had shared with him the notes and sketches she had made during the seminar, he said, "I was so excited about her reaction. Nothing in my many years of teaching calculus had prepared me for this revelation." This experience prompted Professor Leithold to record Dr. Hayman's remarks and share them with other mathematics educators at national and regional meetings. The enthusiasm generated by these presentations led to the publication of THE CALCULUS VIRGIN.

**AN ARTIST'S VIEW OF THE LANGUAGE OF CALCULUS** is revealed by fifty three drawings and texts by Dr. Hayman. This view is derived from her multi-cultural perspective gained from extensive world travel, the unique insight and cosmic scope she possesses as an artist, and what she perceives as "the great poetry and mystical quality of mathematics."

---

**TORTUE PUBLICATIONS**
**2917 SANTA MONICA BLVD.**
**SANTA MONICA, CA. 90404**
**(310) 828-8878**

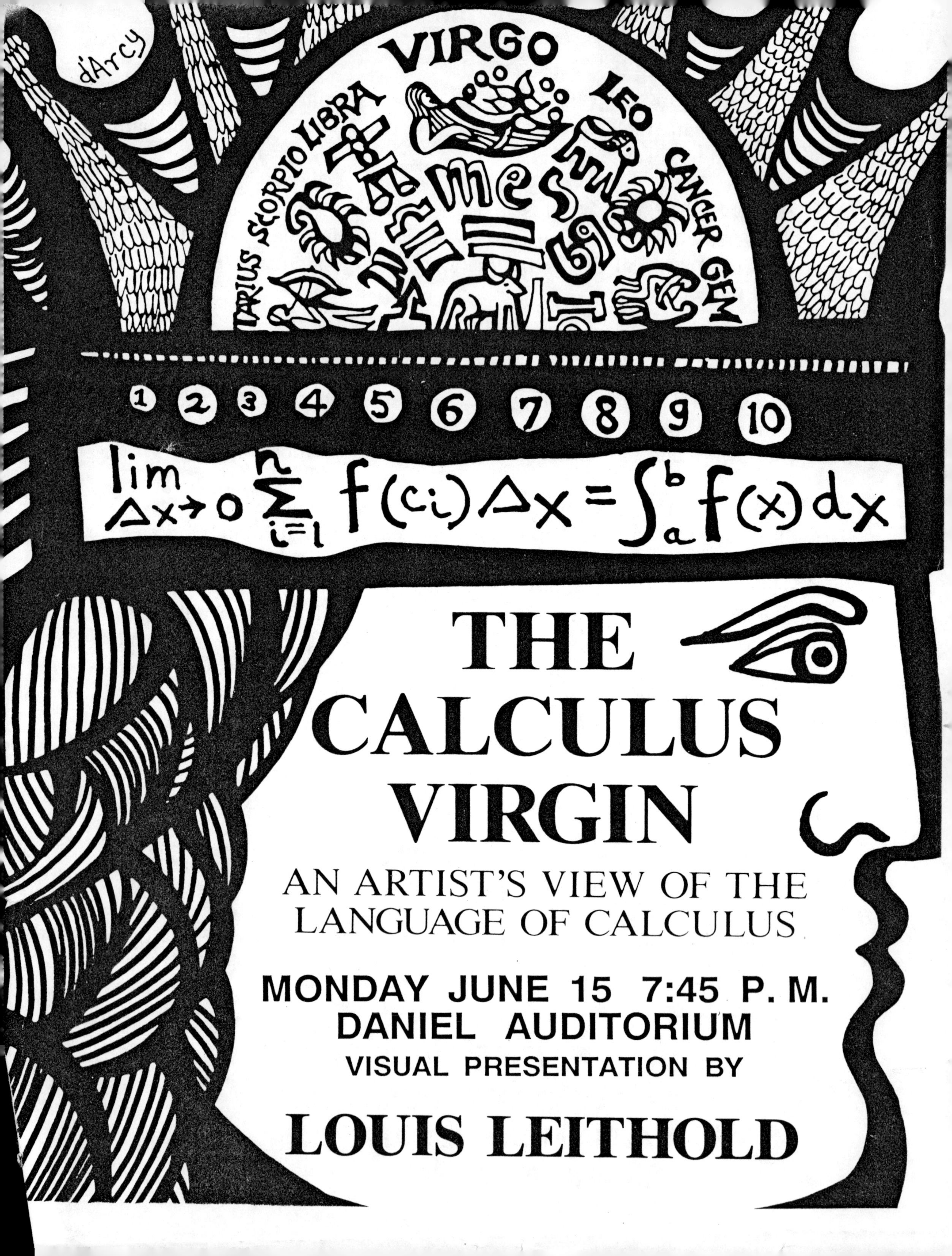
d'Arcy
VIRGO
LEO
CANCER
GEM
LIBRA
SCORPIO
TARIUS
1 2 3 4 5 6 7 8 9 10
$\lim_{\Delta x \to 0} \sum_{i=1}^{n} f(c_i)\Delta x = \int_a^b f(x)\,dx$
THE
CALCULUS
VIRGIN
AN ARTIST'S VIEW OF THE
LANGUAGE OF CALCULUS
MONDAY JUNE 15 7:45 P. M.
DANIEL AUDITORIUM
VISUAL PRESENTATION BY
LOUIS LEITHOLD